Updated Student Edition

**Grade K
Module 6**

Eureka Math™
A Story of Units

Special thanks go to the Gordan A. Cain Center and to the Department of Mathematics at Louisiana State University for their support in the development of *Eureka Math*.

Published by Common Core

Copyright © 2014 Common Core, Inc. All rights reserved. No part of this work may be reproduced or used in any form or by any means – graphic, electronic, or mechanical, including photocopying or information storage and retrieval systems – without written permission from the copyright holder. "Common Core" and "Common Core, Inc.," are registered trademarks of Common Core, Inc.

Common Core, Inc. is not affiliated with the Common Core State Standards Initiative.

Printed in the U.S.A.
This book may be purchased from the publisher at greatminds.net
10 9 8 7 6 5 4 3 2

ISBN 978-1-63255-005-7

Name _____ Date _____

Listen to the directions.

First, draw the missing line to finish the triangle using a ruler. **Second**, color the corners red. **Third**, draw another triangle.

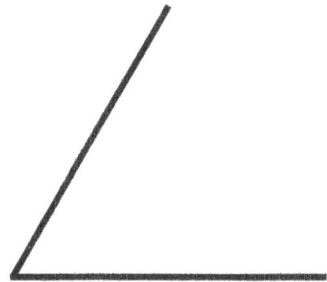

First, use your ruler to draw 2 lines to make a square. **Second**, color the corners red. **Third**, draw another square.

First, draw a triangle using your ruler. **Second,** draw a different triangle using your ruler. **Third**, show your pictures to your partner.

Lesson 1 Problem Set

4 + 1 = ____	5 - 1 = ____
____ = 2 + 1	____ = 4 - 1
3 + 2 = ____	3 - 2 = ____
3 + 1 = ____	3 - 0 = ____
____ = 5 + 0	____ = 5 - 4
2 - 1 = ____	2 + 2 = ____
____ = 3 - 3	____ = 5 - 3
1 - 0 = ____	1 + 1 = ____
3 - 0 = ____	4 - 0 = ____
____ = 4 - 4	____ = 4 + 1

Lesson 1: Describe the systematic construction of flat shapes using ordinal numbers.

A STORY OF UNITS — Lesson 1 Exit Ticket — K•6

Name _____ Date _____

Use your ruler.

First, draw a straight line from the dot.

Second, draw a different straight line from the dot.

Third, draw another straight line to make a triangle.

A STORY OF UNITS   Lesson 1 Homework K•6

Name _____   Date _____

Follow the directions.

First, use your ruler to draw a line finishing the triangle.

Second, color the triangle green.

Third, use your ruler to draw a bigger triangle next to the green triangle.

First, draw 2 lines to make a rectangle.

Second, circle all the corners with a red crayon.

Third, put an X on the longer sides.

First, draw a line to complete the hexagon.

Second, color the hexagon blue.

Third, write the number of sides the hexagon has in the box below.

On the back of your paper, draw:
- A closed shape with 3 straight sides.
- A closed shape with 4 straight sides.
- A closed shape with 6 straight sides.

Lesson 1: Describe the systematic construction of flat shapes using ordinal numbers.

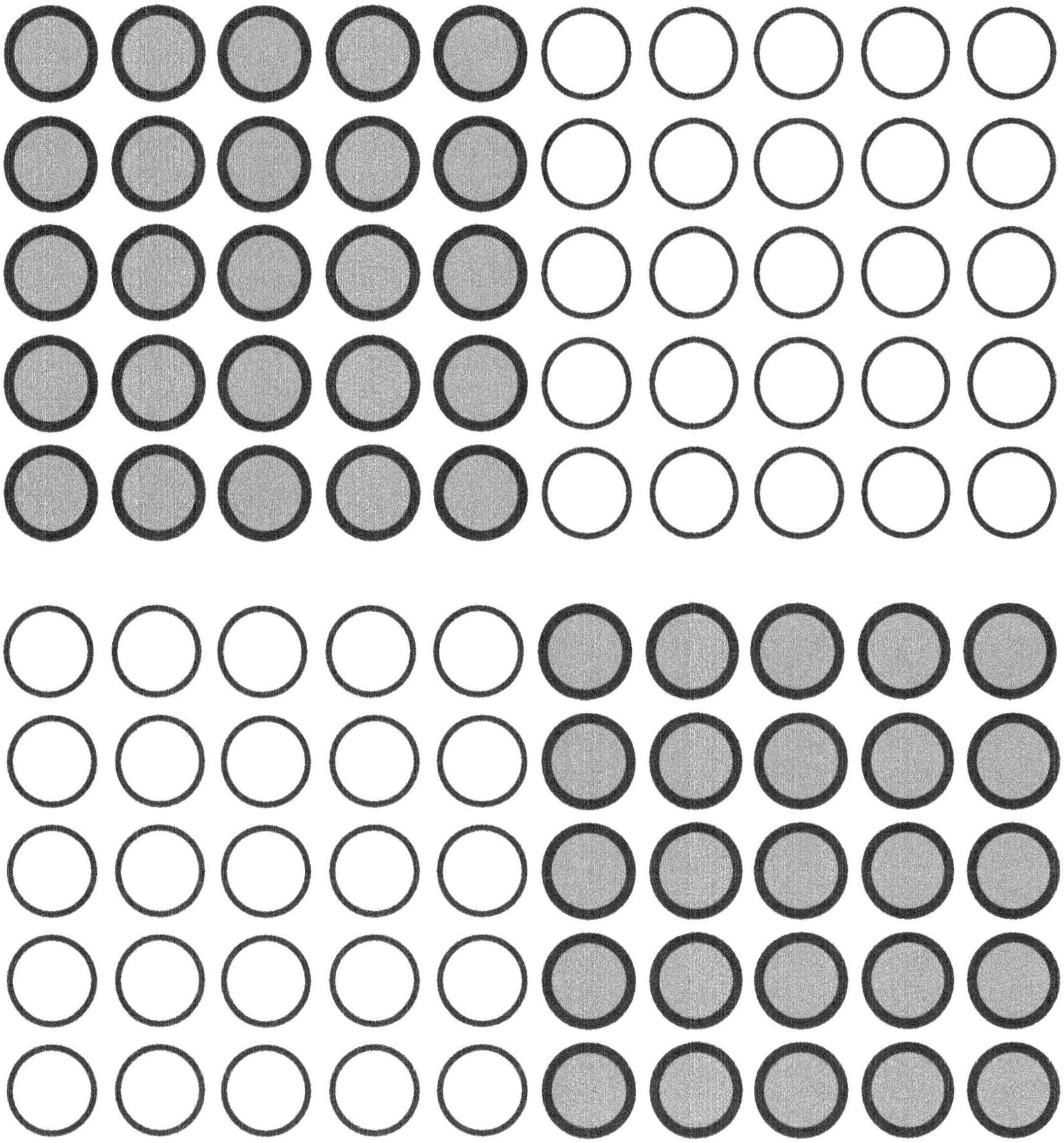

A STORY OF UNITS   Lesson 1 Fluency Template 1   K•6

rekenrek dot paper

Lesson 1:   Describe the systematic construction of flat shapes using ordinal numbers.

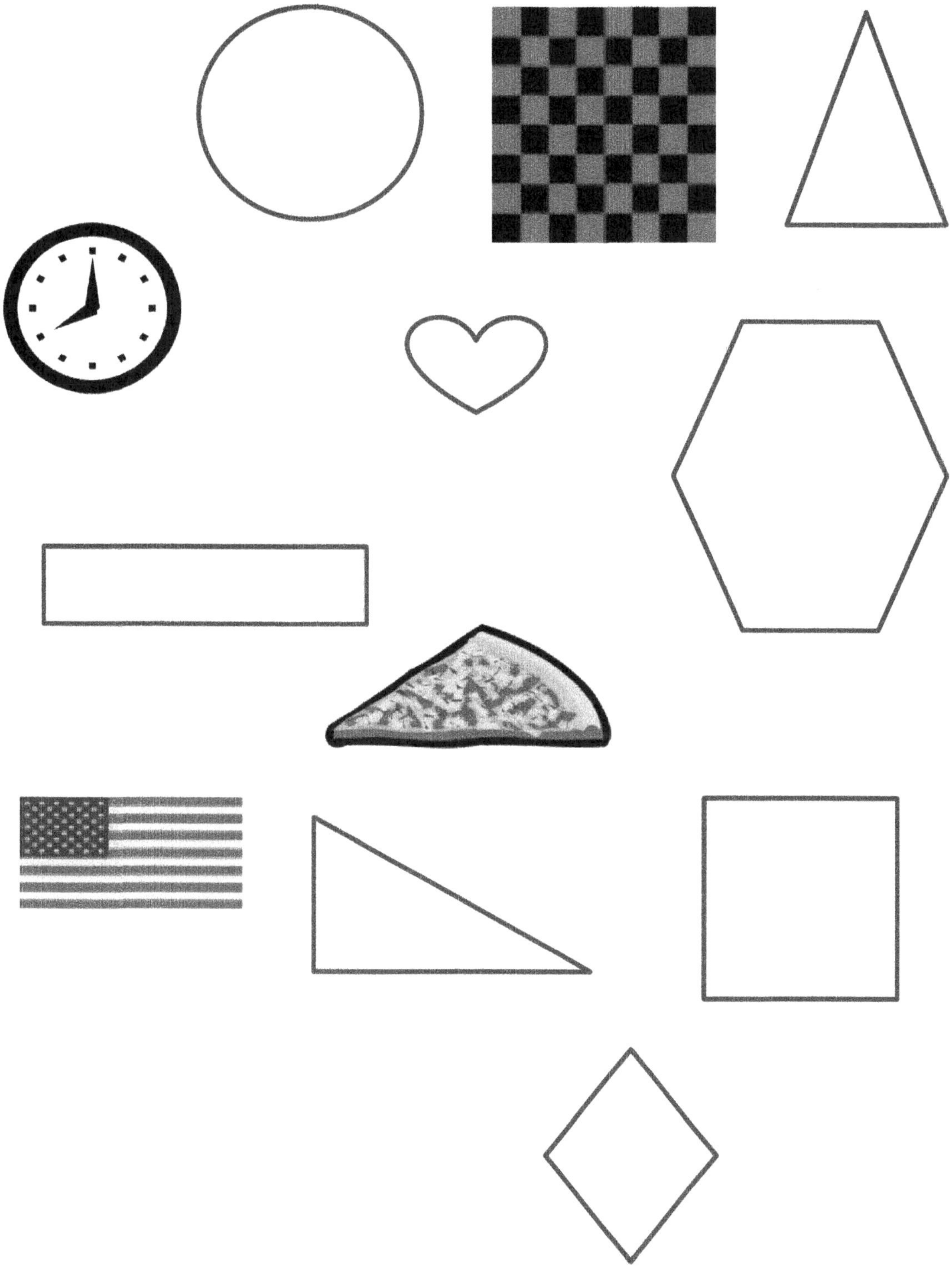

shapes of triangles, rectangles, squares, hexagons, and circles

Lesson 1: Describe the systematic construction of flat shapes using ordinal numbers.

A STORY OF UNITS Lesson 2 Core Fluency Sprint A K•6

Number Correct: ☆

Name _____ Date _____

Write the missing number.

1.	2 + 1 = ☐	11.	☐ = 3 + 2
2.	1 + 1 = ☐	12.	1 + 3 = ☐
3.	1 + 4 = ☐	13.	☐ = 2 + 2
4.	3 + 1 = ☐	14.	☐ = 1 + 2
5.	2 + 2 = ☐	15.	1 + 4 = ☐
6.	2 + 3 = ☐	16.	☐ = 2 + 3
7.	1 + 2 = ☐	17.	☐ = 5 + 1
8.	4 + 1 = ☐	18.	5 + 2 = ☐
9.	3 + 2 = ☐	19.	1 + 0 = ☐
10.	1 + 3 = ☐	20.	5 + 0 = ☐

Lesson 2: Build flat shapes with varying side lengths and record with drawings.

A STORY OF UNITS — Lesson 2 Core Fluency Sprint B — K•6

Number Correct:

Name _____ Date _____

Write the missing number.

1.	2 - 1 = ☐	11.	☐ = 4 - 2
2.	4 - 1 = ☐	12.	5 - 3 = ☐
3.	5 - 1 = ☐	13.	☐ = 3 - 1
4.	3 - 1 = ☐	14.	☐ = 5 - 2
5.	3 - 2 = ☐	15.	4 - 1 = ☐
6.	4 - 2 = ☐	16.	☐ = 5 - 4
7.	5 - 3 = ☐	17.	☐ = 5 - 1
8.	5 - 2 = ☐	18.	6 - 1 = ☐
9.	4 - 3 = ☐	19.	1 - 0 = ☐
10.	5 - 4 = ☐	20.	5 - 5 = ☐

Lesson 2: Build flat shapes with varying side lengths and record with drawings.

A STORY OF UNITS Lesson 2 Core Fluency Sprint C K•6

Name _____ Date _____

Number Correct: _____

Write the missing number.

1.	2 + 1 = ☐	11.	3 + 2 = ☐
2.	2 - 1 = ☐	12.	3 - 2 = ☐
3.	3 + 1 = ☐	13.	4 + 0 = ☐
4.	3 - 1 = ☐	14.	4 - 0 = ☐
5.	4 + 1 = ☐	15.	5 + 0 = ☐
6.	4 - 1 = ☐	16.	5 - 0 = ☐
7.	1 + 1 = ☐	17.	5 - 5 = ☐
8.	1 - 1 = ☐	18.	4 + 1 = ☐
9.	2 + 2 = ☐	19.	5 - 4 = ☐
10.	2 - 2 = ☐	20.	5 - 1 = ☐

EUREKA MATH

Lesson 2: Build flat shapes with varying side lengths and record with drawings.

A STORY OF UNITS　　　　　　　　　　　　　　Lesson 2 Core Fluency Sprint D　K•6

Number Correct: _____

Name _____　Date _____

Write the missing number.

1.	2 + 1 = ☐		11.	☐ = 1 + 2
2.	4 + 1 = ☐		12.	5 + 0 = ☐
3.	5 − 1 = ☐		13.	☐ = 3 − 1
4.	3 + 1 = ☐		14.	☐ = 2 + 2
5.	3 + 2 = ☐		15.	4 − 1 = ☐
6.	4 − 2 = ☐		16.	☐ = 5 − 4
7.	5 − 3 = ☐		17.	☐ = 5 − 1
8.	5 − 2 = ☐		18.	3 + 0 = ☐
9.	2 + 3 = ☐		19.	1 − 0 = ☐
10.	5 − 4 = ☐		20.	5 − 5 = ☐

Lesson 2:　Build flat shapes with varying side lengths and record with drawings.

A STORY OF UNITS

Lesson 2 Problem Set K•6

Name _____ Date _____

First, use a ruler to trace the shapes. Second, follow the directions in each box. Use your ruler to draw the shapes.

Draw 3 different triangles.

Draw 2 different rectangles.

Draw 1 hexagon.

Lesson 2: Build flat shapes with varying side lengths and record with drawings.

5 - 4 = ____ 5 - 3 = ____ 5 - 2 = ____ 5 - 1 = ____ 5 - 0 = ____	0 + 1 = ____ 1 + 1 = ____ 2 + 1 = ____ 3 + 1 = ____ 4 + 1 = ____
4 - 2 = ____ 2 - 1 = ____ 3 - 2 = ____ 3 - 1 = ____ 5 - 0 = ____	4 - 3 = ____ 2 + 1 = ____ 3 + 2 = ____ 4 - 1 = ____ 5 - 4 = ____

Lesson 2: Build flat shapes with varying side lengths and record with drawings.

A STORY OF UNITS   Lesson 2 Exit Ticket K•6

Name _____ Date _____

First, draw a triangle so all of the sides are different lengths.

Second, draw a triangle with your ruler that has 2 sides that are about the same length.

Lesson 2: Build flat shapes with varying side lengths and record with drawings.

A STORY OF UNITS
Lesson 2 Homework K•6

Name _____ Date _____

Trace the shapes. Then, use a ruler to draw similar shapes, on your own, in the large rectangle. Draw more on the back of your paper if you would like!

Lesson 2: Build flat shapes with varying side lengths and record with drawings.

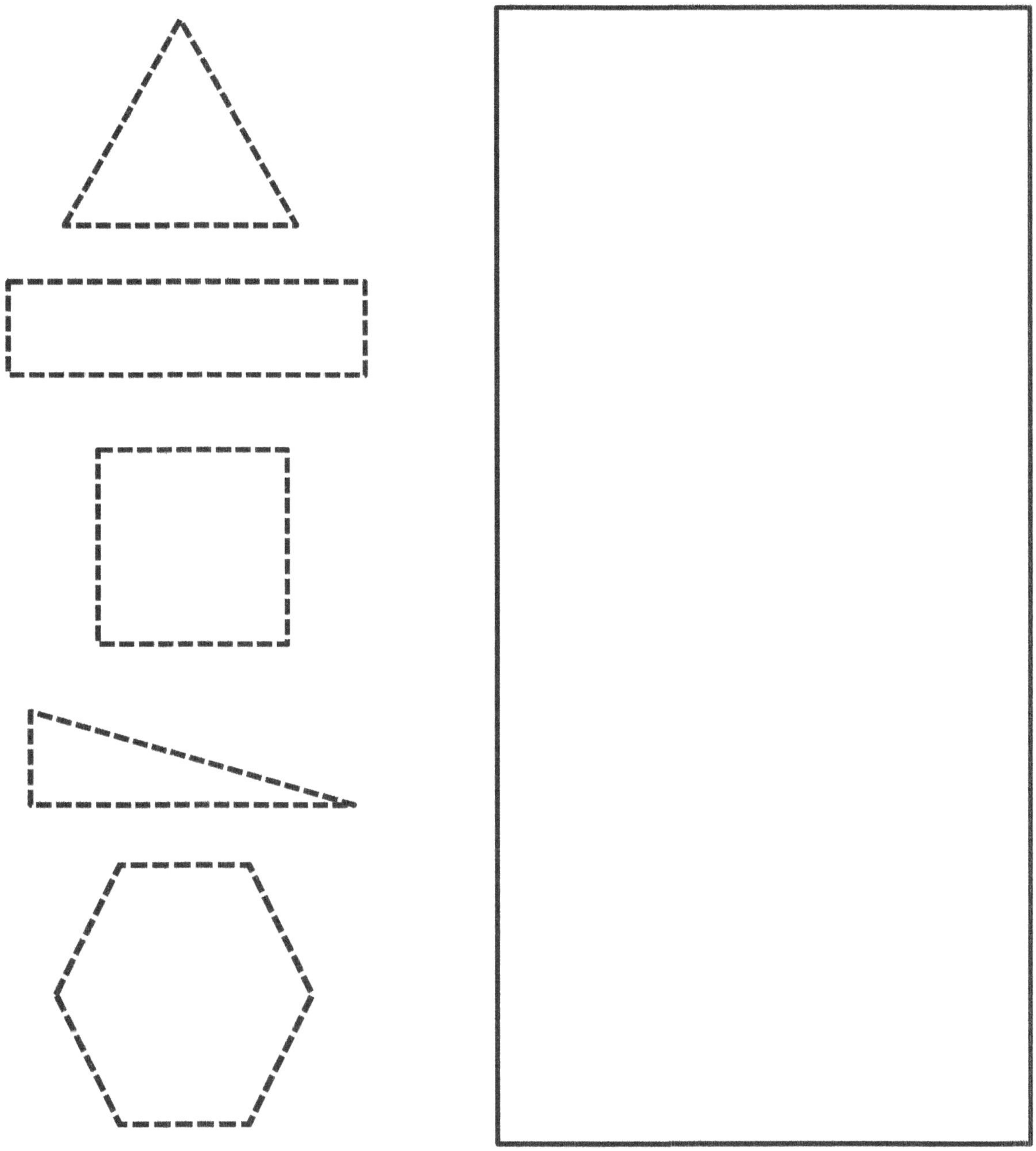

A STORY OF UNITS   Lesson 2 Homework K•6

Lesson 2:   Build flat shapes with varying side lengths and record with drawings.

Numerals

1	0		
0	1	2	3
4	5	6̲	7
8	9̲		

hide zero cards (numeral side) (Copy double-sided on cardstock with 5-groups page, and cut.)

Lesson 2: Build flat shapes with varying side lengths and record with drawings.

A STORY OF UNITS — Lesson 2 Fluency Template — K•6

5-Groups

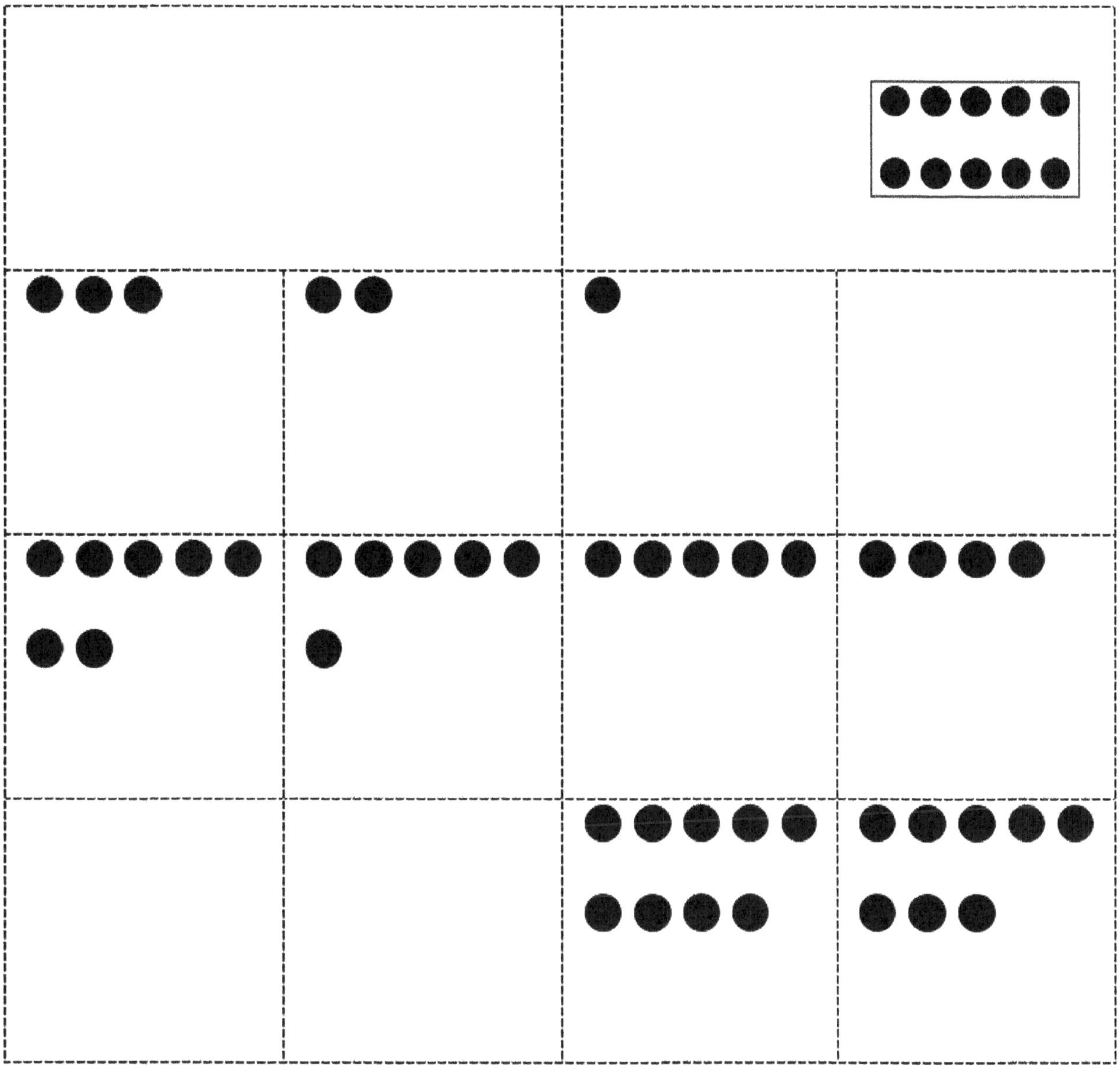

hide zero cards (5-groups side) (Copy double-sided on cardstock with numerals page, and cut.)

Lesson 2: Build flat shapes with varying side lengths and record with drawings.

Name _____ Date _____

Trace the circles and rectangle. Cut out the shape. Fold and tape to create a cylinder.

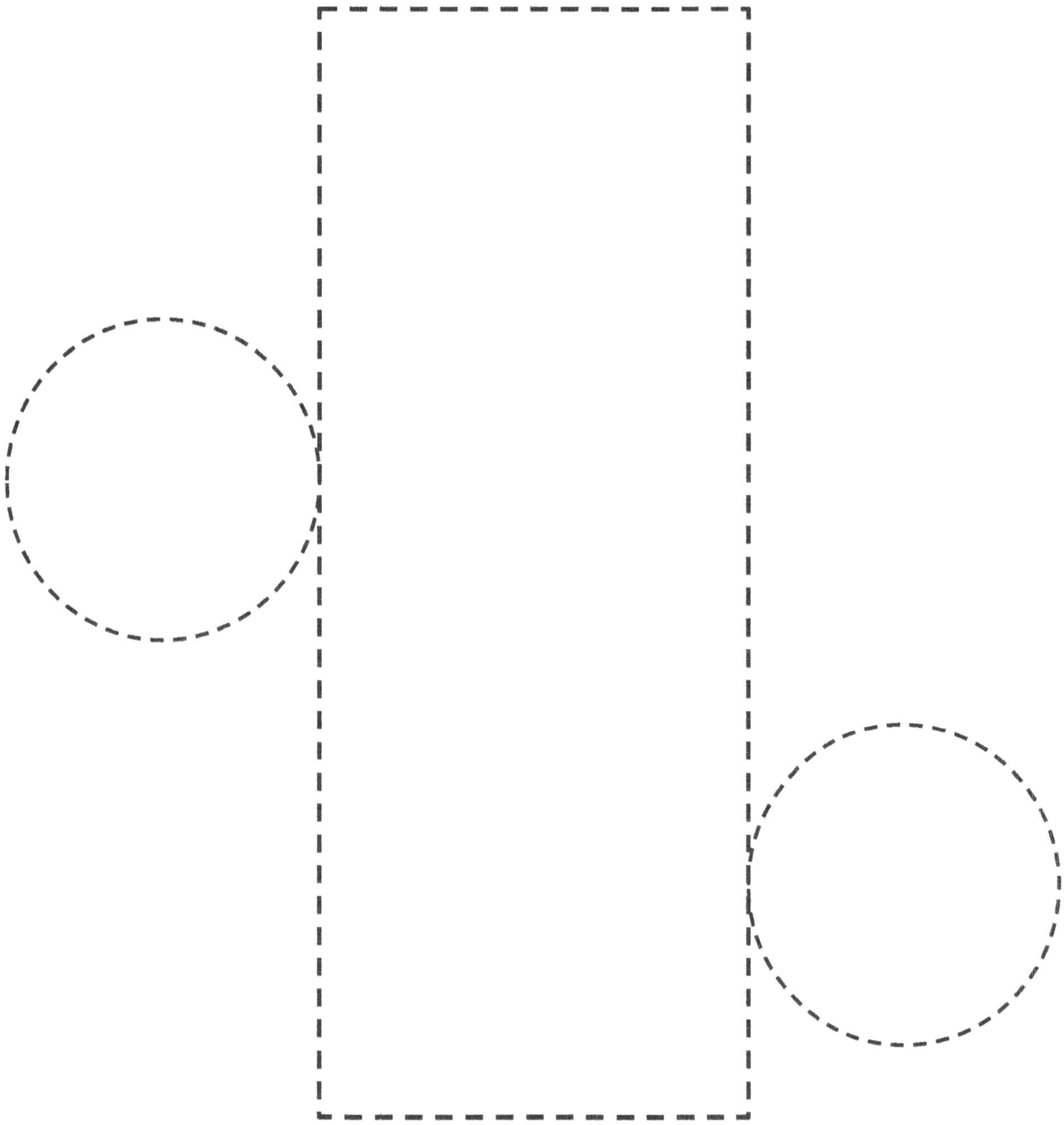

Lesson 3: Compose solids using flat shapes as a foundation.

Trace the squares. Cut out the shape. Fold and tape to create a cube.

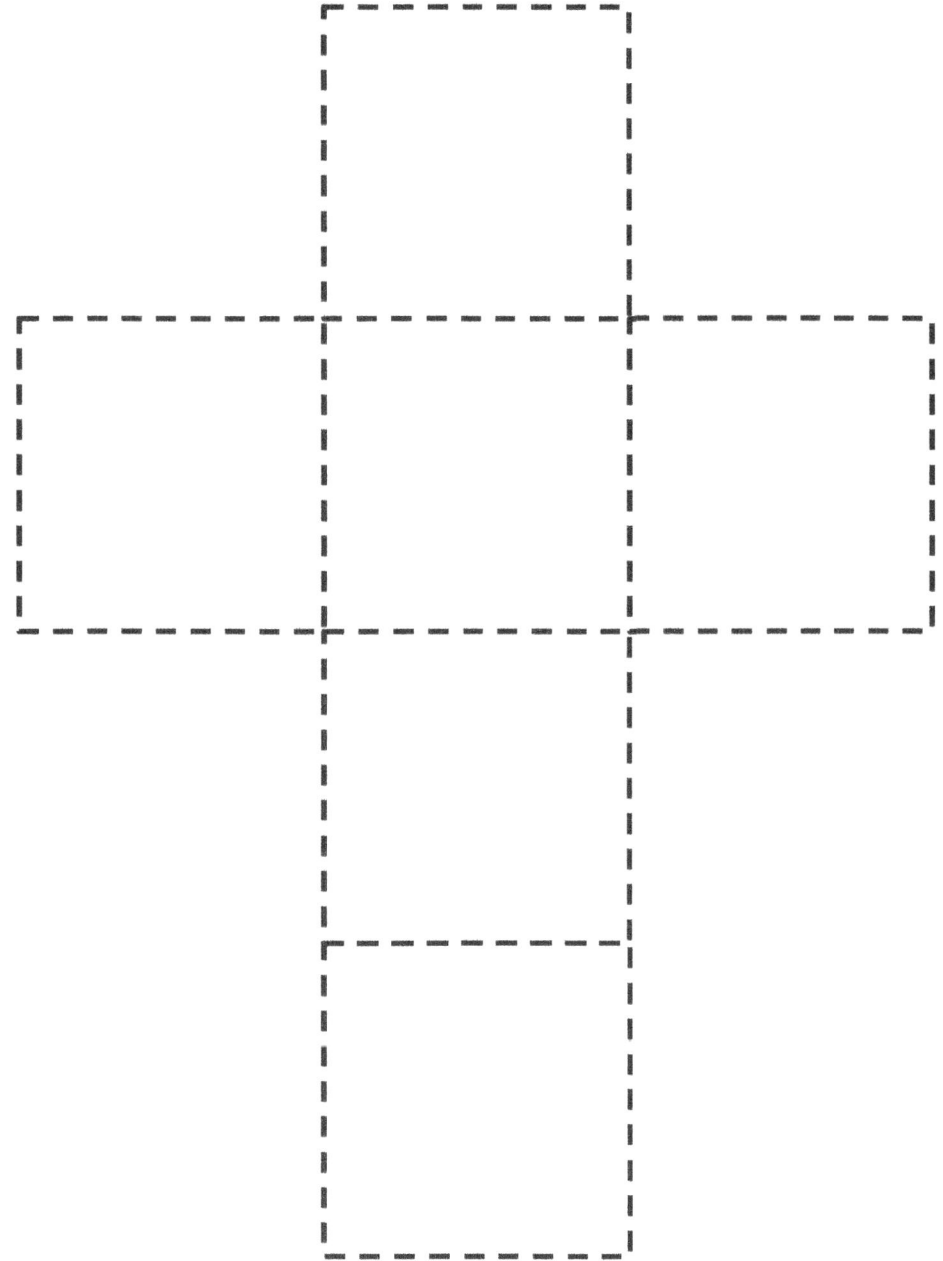

Name _____   Date _____

Draw a line from the flat shape to the object that has a face with that flat shape.

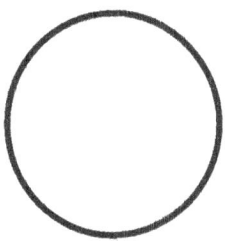   

Lesson 3: Compose solids using flat shapes as a foundation.

A STORY OF UNITS   Lesson 3 Homework   K•6

Name _____ Date _____

Draw something that is a cylinder.

Circle the flat shape you can see in a ⌭.   □ ○

Draw something that is a cube.

Circle the flat shape you can see in a ▣.   □ ⬡

Lesson 3:   Compose solids using flat shapes as a foundation.

A STORY OF UNITS   Lesson 3 Homework K•6

Draw something that is a cone.

Circle the flat shape you can see in a .   △   ○

Draw a 3-dimensional solid. Draw one of your solid's faces. Tell an adult about the shapes you drew.

Note to Family Helpers: Your child knows how to name some 3-dimensional solids: cylinders, cones, cubes, and spheres. You can often find these 3-D shapes around the house in objects such as soup cans, ice cream cones, boxes, and balls. For the last question, it is acceptable for your student to find and draw a different type of 3-D solid. Talk about the number of edges, corners, and faces on the object.

 Lesson 3: Compose solids using flat shapes as a foundation.

A STORY OF UNITS   Lesson 3 Fluency Template 1   K•6

Name _____   Date _____

Add. Color the blocks using the code for the total.

1—RED	2—ORANGE	3—YELLOW
4—GREEN	5—BLUE	

0 + 1	1 + 1	2 + 1	3 + 1	4 + 1
0 + 2	1 + 2	2 + 2	3 + 2	
0 + 3	1 + 3	2 + 3		
0 + 4	1 + 4			
0 + 5				

color by answer addition

Lesson 3: Compose solids using flat shapes as a foundation.

A STORY OF UNITS   Lesson 3 Fluency Template 2   K•6

Name _____   Date _____

Subtract. Color the blocks using the code for the difference.

| 0—PURPLE | 1—RED | 2—ORANGE | 3—YELLOW |
| 4—GREEN | 5—BLUE | | |

1 - 0	2 - 0	3 - 0	4 - 0	5 - 0
1 - 1	2 - 1	3 - 1	4 - 1	5 - 1
	2 - 2	3 - 2	4 - 2	5 - 2
		3 - 3	4 - 3	5 - 3
			4 - 4	5 - 4
				5 - 5

color by answer subtraction

Lesson 3: Compose solids using flat shapes as a foundation.

Name _____ Date _____

Circle the 2nd truck from the stop sign. Draw a square around the 5th truck. Draw an X on the 9th truck.

Draw a triangle around the 4th vehicle from the stop sign. Draw a circle around the 1st vehicle. Draw a square around the 6th vehicle.

Put an X on the 10th horse from the stop sign. Draw a triangle around the 7th horse. Draw a circle around the 3rd horse. Draw a square around the 8th horse.

Lesson 4: Describe the relative position of shapes using ordinal numbers.

Draw a line from the shape to the correct ordinal number, starting at the top.

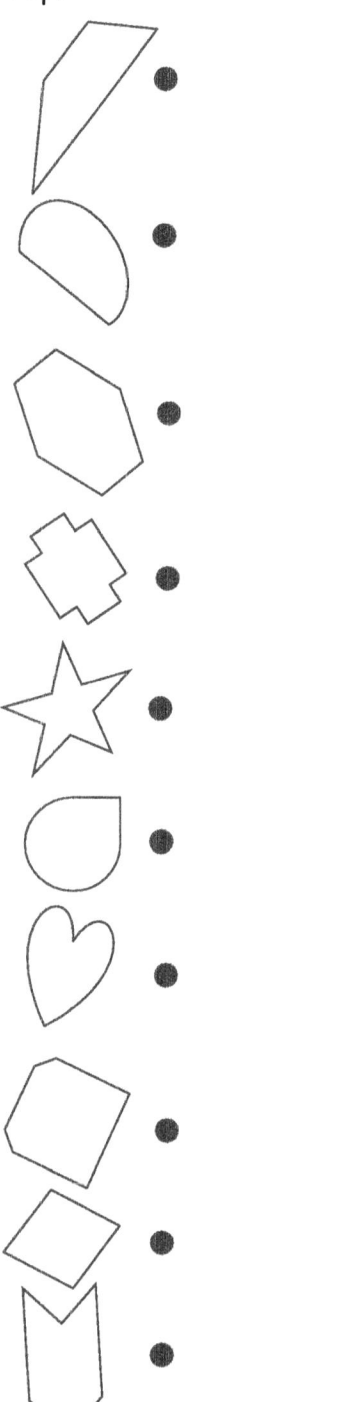

	9th ninth
	4th fourth
	6th sixth
	1st first
	7th seventh
	3rd third
	10th tenth
	5th fifth
	8th eighth
	2nd second

Lesson 4: Describe the relative position of shapes using ordinal numbers.

Name _____ Date _____

Listen to the directions. Start at the circle when counting.

Color the 5th shape red.
Color the 2nd shape green.
Color the 10th shape yellow.
Color the 7th shape blue.
Color the 1st shape pink.
Color the 8th shape orange.

Name _____   Date _____

Color the 1ˢᵗ ☆ red.
Color the 3ʳᵈ ☆ blue.
Color the 5ᵗʰ ☆ green.
Color the 8ᵗʰ ☆ purple.

Put an X on the 2ⁿᵈ shape.
Draw a triangle in the 4ᵗʰ shape.
Draw a circle around the 6ᵗʰ shape.
Draw a square in the 9ᵗʰ shape.

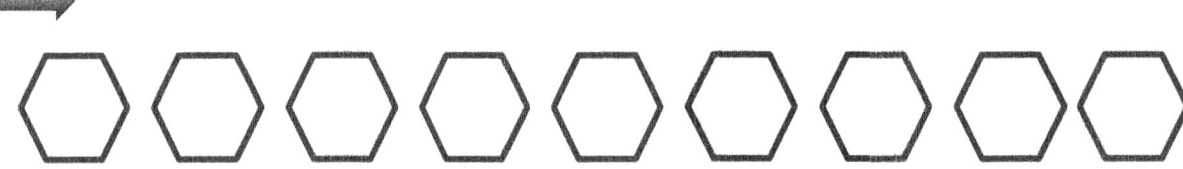

Draw a circle in the 7ᵗʰ shape.
Put an X on the 1ˢᵗ shape.
Draw a square in the 5ᵗʰ shape.
Draw a triangle in the 3ʳᵈ shape.

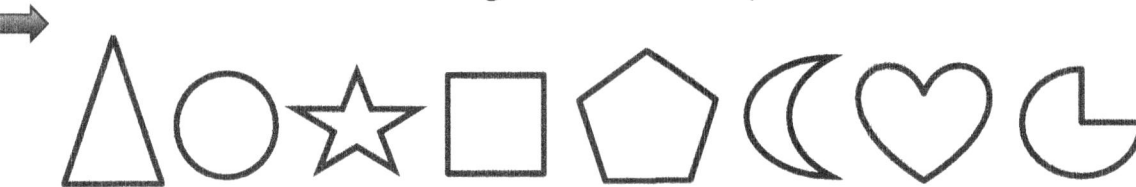

Lesson 4: Describe the relative position of shapes using ordinal numbers.

Match each animal to the place where it finished the race.

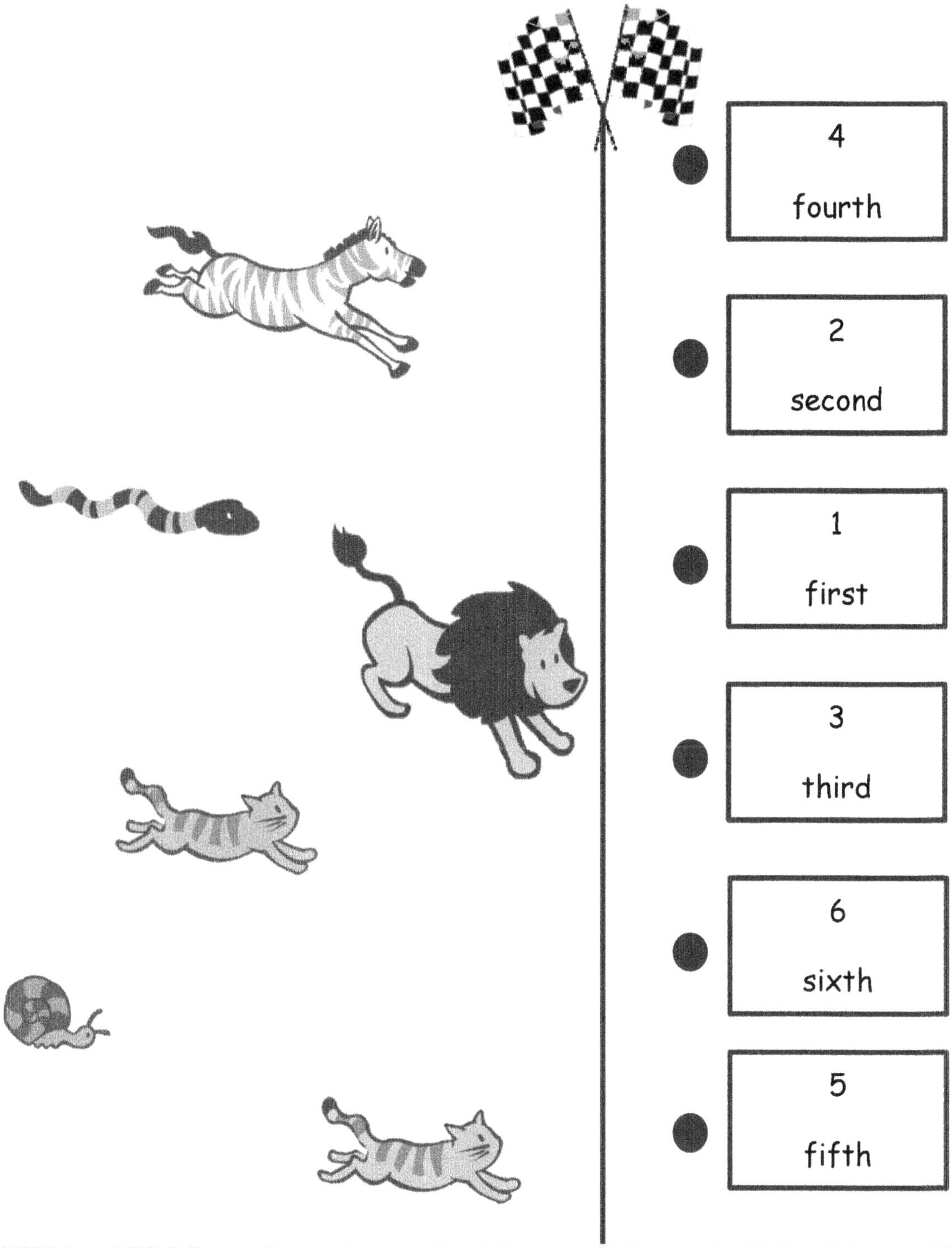

	4 fourth
Zebra	2 second
Snake	
Lion	1 first
	3 third
Cat (small)	
Snail	6 sixth
Cat (striped)	5 fifth

Lesson 4: Describe the relative position of shapes using ordinal numbers.

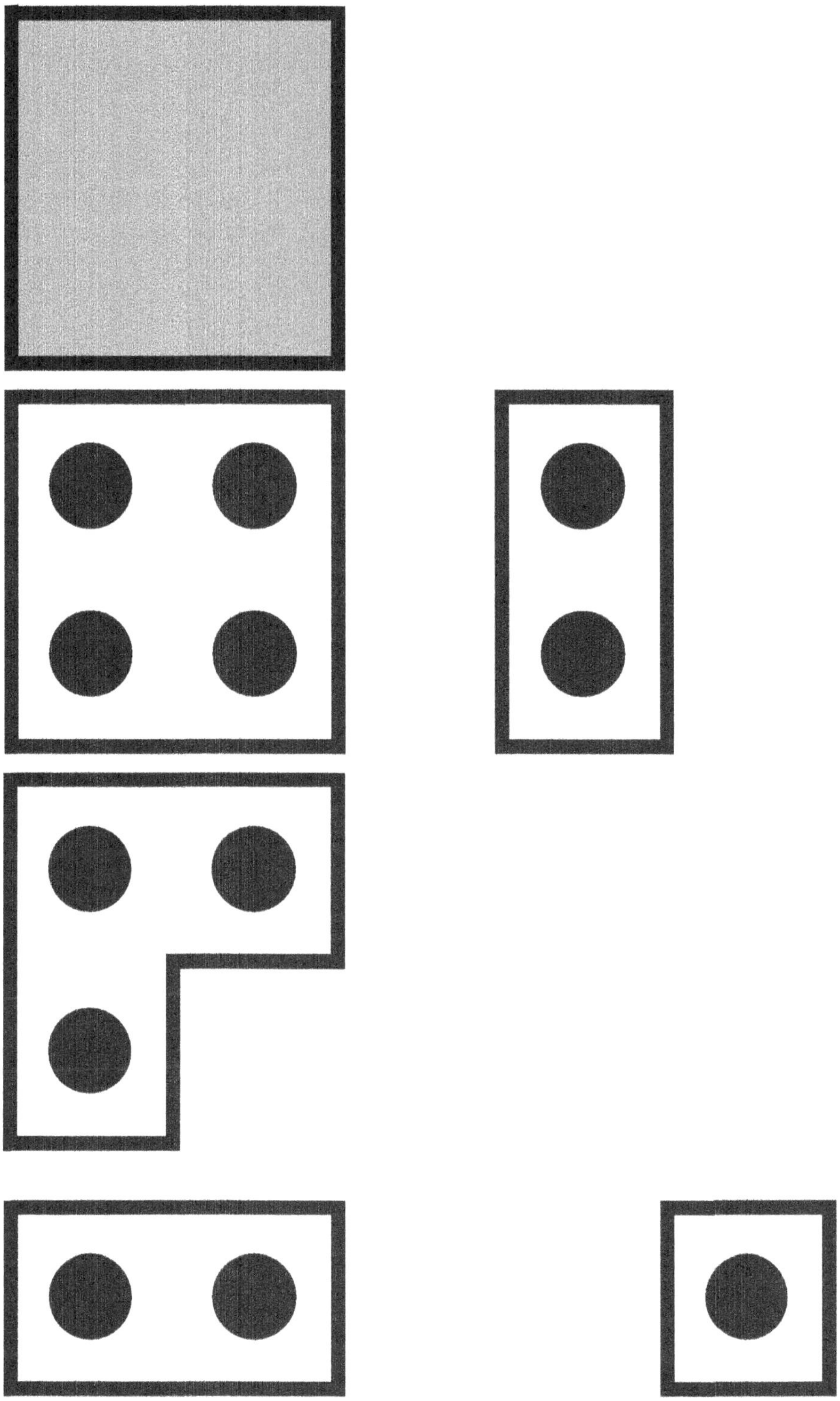

A STORY OF UNITS — Lesson 4 Fluency Template 1 K•6

4-dot puzzle cards

Lesson 4: Describe the relative position of shapes using ordinal numbers.

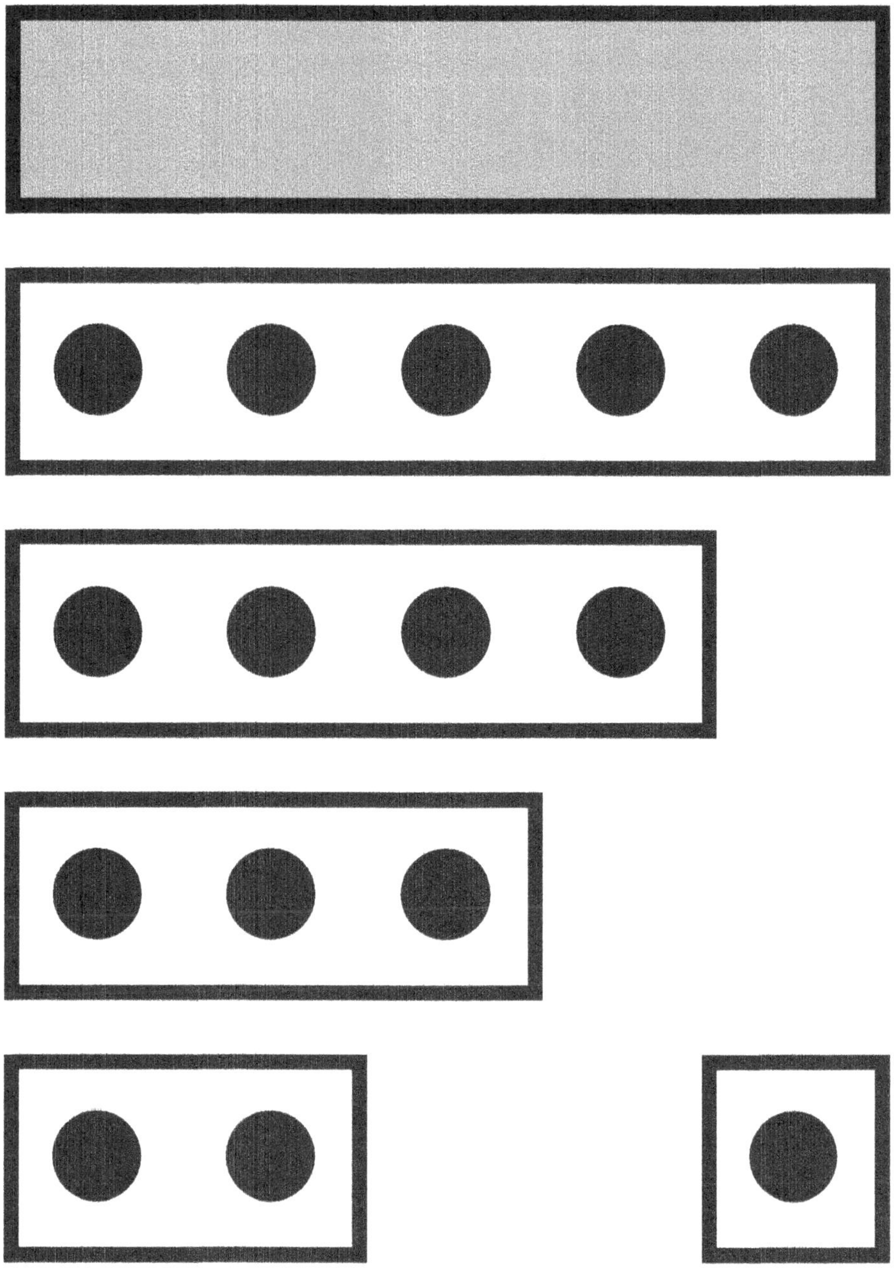

A STORY OF UNITS

Lesson 4 Fluency Template 2   K•6

5-dot puzzle cards

EUREKA MATH

Lesson 4: Describe the relative position of shapes using ordinal numbers.

shapes

Name _____ Date _____

Choose 4 shapes to create a new shape in Box 1. Give the same 4 shapes to your partner. Have your partner create a different shape in Box 2.

1

2

Lesson 5: Compose flat shapes using pattern blocks and drawings.

A STORY OF UNITS

Lesson 5 Problem Set K•6

Choose 5 shapes to create a new shape in Box 3. Give the same 5 shapes to your partner. Have your partner create a different shape in Box 4.

3

4

Subtract.

5 − 1 = ☐   5 − 2 = ☐   5 − 3 = ☐   5 − 4 = ☐

Lesson 5: Compose flat shapes using pattern blocks and drawings.

Name _____ Date _____

Use your pattern blocks to help you solve the problem.

Use 2 blocks to make a rectangle. Trace your blocks to show your rectangle.

Name _____ Date _____

Match each group of shapes on the left with the new shape they make when they are put together.

 •   •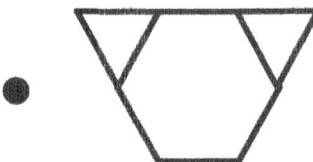

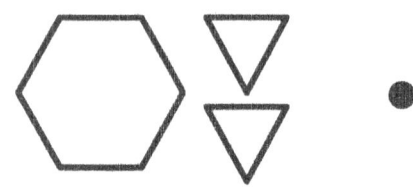 •   •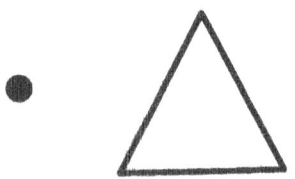

 •   •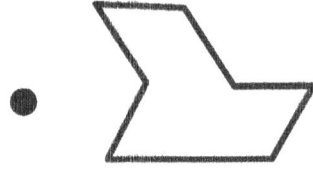

 •   •

Lesson 5: Compose flat shapes using pattern blocks and drawings.

I Can Make New Shapes!

I can make new shapes recording sheet

Lesson 5: Compose flat shapes using pattern blocks and drawings.

Number Correct:

Name ______________________ Date ____________

Write the number of dots needed to make 10 dots.

1.	●●●●● ●●●●	9.	●
2.	●●●●● ●●●	10.	●●●●● ●●●●
3.	●●●●● ●●	11.	●●●●● ●●●
4.	●●●●● ●	12.	●●
5.	●●●●●	13.	●●●●● ●●
6.	●●●●	14.	●●●
7.	●●●	15.	●●●●● ●
8.	●●	16.	●●●●

Name _____ Date _____

Trace to show 2 ways to make each shape. How many shapes did you use?

I used __3__ shapes.   I used _____ shapes.

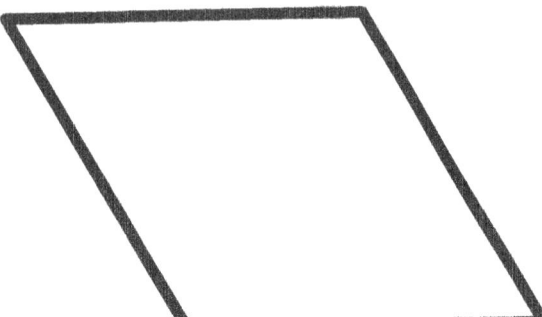 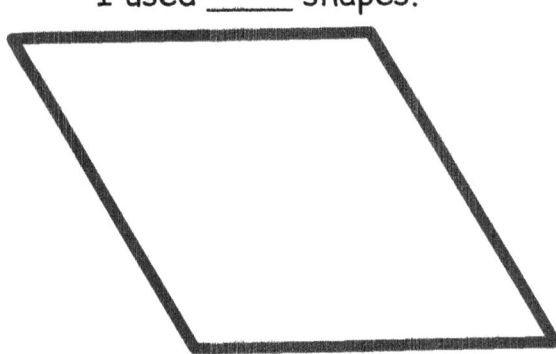

I used _____ shapes.   I used _____ shapes.

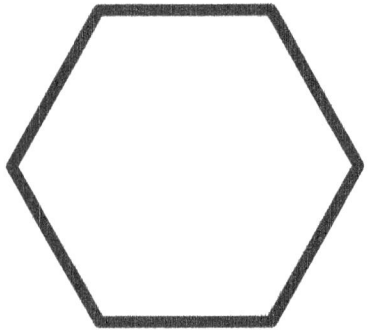 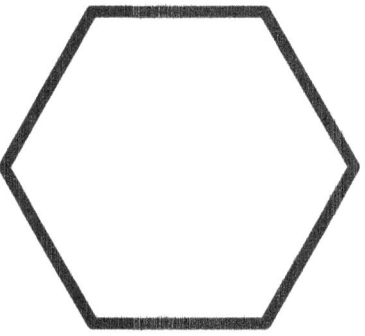

I used _____ shapes.   I used _____ shapes.

Lesson 6:   Decompose flat shapes into two or more shapes.

A STORY OF UNITS

Lesson 6 Problem Set K•6

Fill in each shape with pattern blocks. Trace to show the shapes you used.

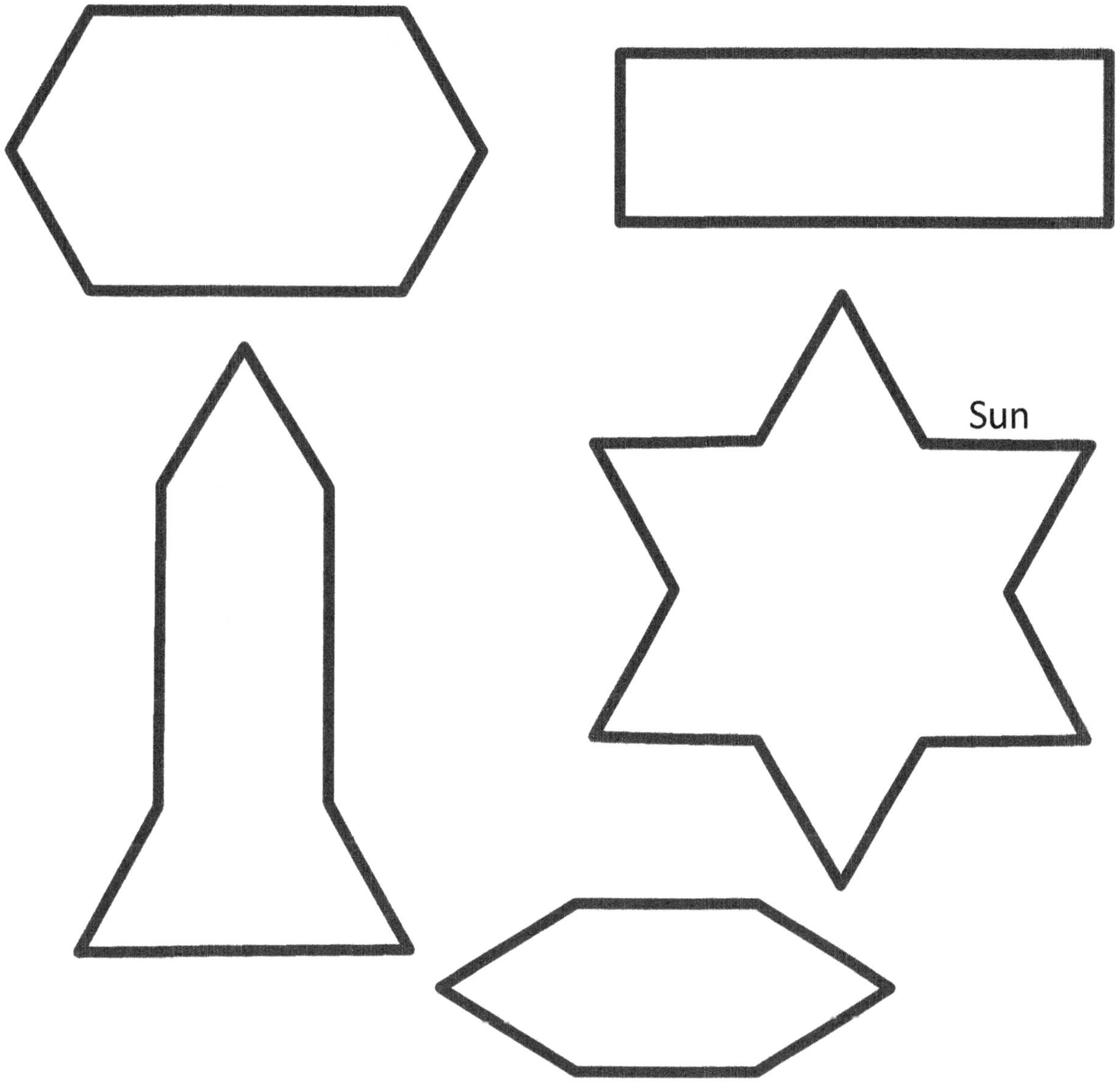

How many different ways can you cover the sun picture with pattern blocks?

Answer: 3 ways
(2 triangles), (6 triangles/1 hexagon), (1 large triangle/3 small triangles)

Lesson 6:   Decompose flat shapes into two or more shapes.

A STORY OF UNITS — Lesson 6 Exit Ticket K•6

Name _____ Date _____

Draw 2 shapes that can be used to build the rectangle.

Draw 2 shapes that can be used to build the house.

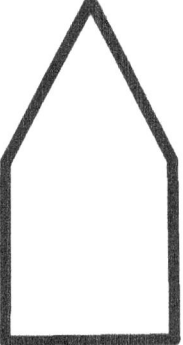

Lesson 6: Decompose flat shapes into two or more shapes.

A STORY OF UNITS   Lesson 6 Homework K•6

Name _____ Date _____

Cut out the triangles at the bottom of the paper. Use the small triangles to make the big shapes. Draw lines to show where the triangles fit. Count how many small triangles you used to make the big shapes.

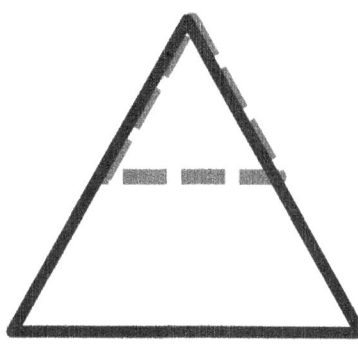

This big triangle is made with ____ small triangles.

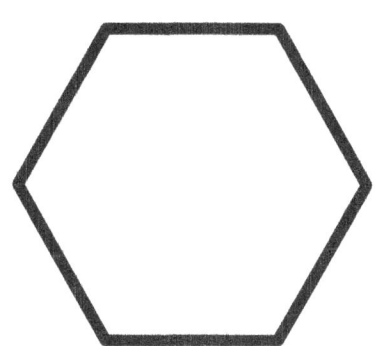

This hexagon is made with ____ small triangles.

- -

△ △ △ △ △ △

Lesson 6: Decompose flat shapes into two or more shapes.

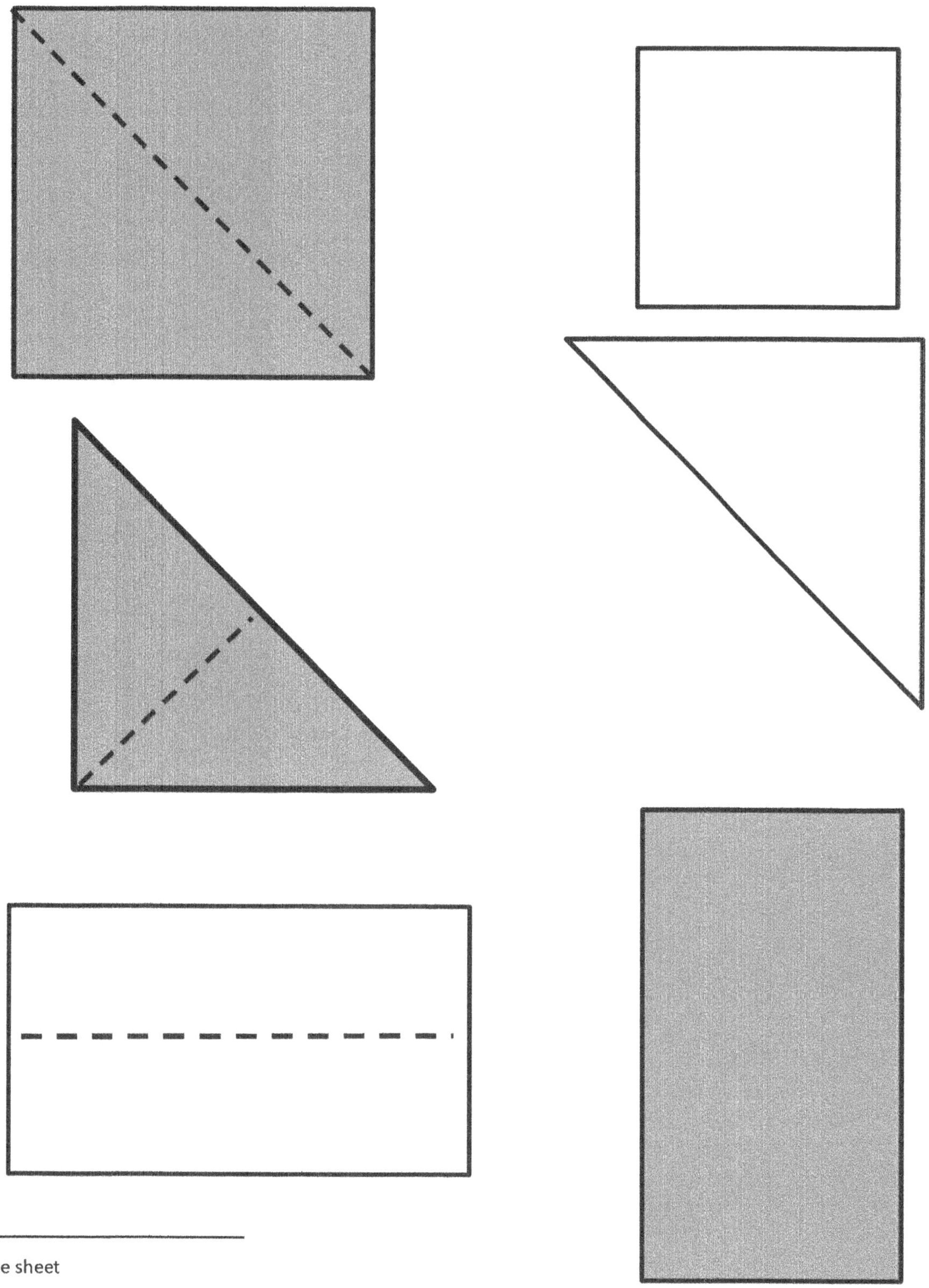

A STORY OF UNITS

Lesson 6 Template K•6

shape sheet

Lesson 6: Decompose flat shapes into two or more shapes.

A STORY OF UNITS   Lesson 7 Problem Set   K•6

Name _____   Date _____

Glue your puzzles into the frames.

[Glue puzzle here.]

[Glue puzzle here.]

Draw some of the shapes that you had after you cut your rectangles.

Lesson 7:   Compose simple shapes to form a larger shape described by an outline.

Carlos drew 2 lines on his square. You can see his square before he cut it. Circle the shapes Carlos had after he cut.

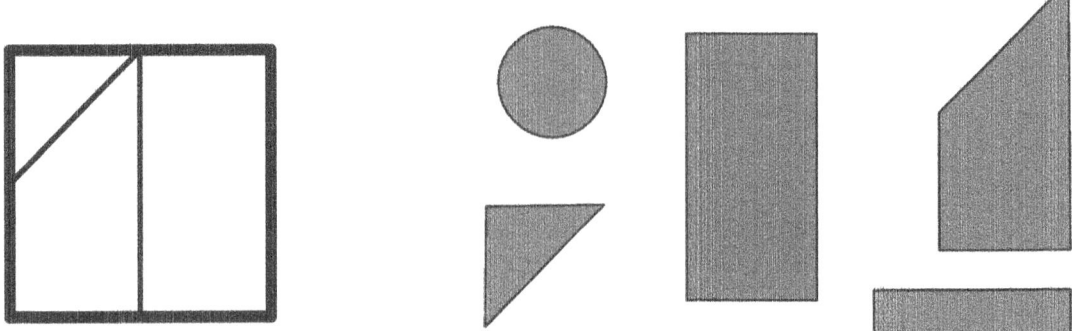

India drew 2 lines on her rectangle. You can see her rectangle before she cut it. Circle the shapes India had after she cut.

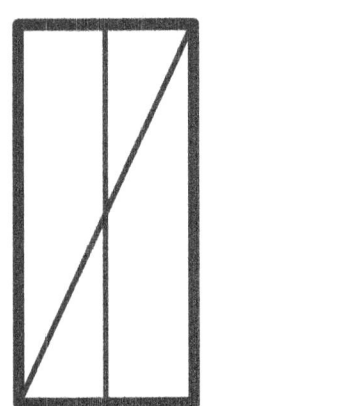 

A STORY OF UNITS

Lesson 7 Exit Ticket K•6

Name _____ Date _____

If you drew 2 straight lines inside the gray rectangle, what shapes might you find? Circle them.

Lesson 7: Compose simple shapes to form a larger shape described by an outline.

A STORY OF UNITS

Lesson 7 Homework K•6

Name _____ Date _____

Using your ruler, draw 2 straight lines from side to side through each shape. The first one has been started for you. Describe to an adult the new shapes you made.

Lesson 7: Compose simple shapes to form a larger shape described by an outline.

A STORY OF UNITS — Lesson 7 Fluency Template — K•6

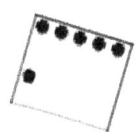
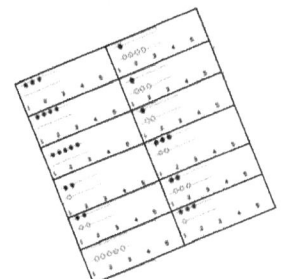

I'm Getting Ready for 1st Grade!

My Math Fluency Kit

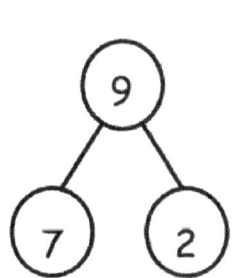
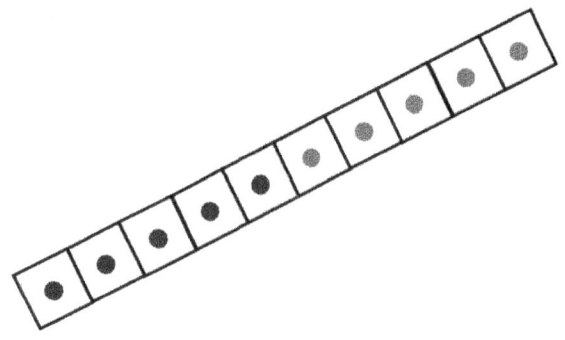

Name _____

fluency kit

Lesson 7: Compose simple shapes to form a larger shape described by an outline.

Name _____   Date _____

My Plan to Get Ready for 1st Grade Math

This is a picture of someone who can help me practice.

This is a picture of where I will practice.

This is ME getting ready for 1st grade!

fluency kit

Name _____

My Sprint Progress Log

Practice your number sentences and Sprints on your personal white board. Ask an adult to time you. Keep track of how you improve over the summer.

Date	Time

Are you getting better at your number sentences?

fluency kit

Lesson 7: Compose simple shapes to form a larger shape described by an outline.

A STORY OF UNITS — Lesson 7 Template K•6

shape puzzle

Lesson 7: Compose simple shapes to form a larger shape described by an outline.

A STORY OF UNITS — Lesson 8 Recording Sheet K•6

Name _____   Date _____

A. **Make 10 Mania:** Show how you made 10.

- -

Name _____   Date _____

B. **Five-Group Frenzy:** Write the number, draw the number in the 5-group way, and draw the number in any other configuration.

Lesson 8: Culminating task—review selected topics to create a cumulative year-end project.

A STORY OF UNITS   Lesson 8 Recording Sheet   K•6

Name _____   Date _____

C. Shape Shifters: Choose 5 pattern blocks, and create a shape. Trace your shape, and then trade with a partner.

- -

Name _____   Date _____

D. The Weigh Station: Choose an object. Guess how many pennies are the same weight as the object. Then, see if you guessed correctly! Draw a picture of the object, and write how many pennies it weighs.

Lesson 8: Culminating task—review selected topics to create a cumulative year-end project.

Name _____ Date _____

E. Awesome Authors: Roll the die. Use the number to create an addition or take-away sentence. Draw a picture, number bond, and number sentence. Share your story with a friend.

A STORY OF UNITS | End-of-Module Assessment Task K•6

Student Name _____

Topic A: Building and Drawing Flat and Solid Shapes

	Date 1	Date 2	Date 3
Topic A			
Topic B			

Rubric Score: _____ Time Elapsed: _____

Materials: (S) 1 set of four 3" straws, 1 set of four 5" straws (separated by length for the students), small clay balls for connectors, 5 real world items with familiar shapes (e.g., book, clock, including a square and rectangle), pattern blocks (Template 1)

1. (Place all straws and formed clay connecting balls in front of the student.) Build a square.
2. (Place solid shapes in front of the student.) Choose one object that has the shape you just built.
3. (Place pattern blocks template in front of the student horizontally.) The star is the beginning. Point to the third shape. Point to the seventh shape.
4. (Turn the template vertically.) The star is the beginning. Point to the first shape. Point to the ninth shape.

What did the student do?	What did the student say?
1.	
2.	
3.	
4.	

Module 6: Analyzing, Comparing, and Composing Shapes

End-of-Module Assessment Task K•6

Topic B: Composing and Decomposing Shapes

Rubric Score: _____ Time Elapsed: _____

Materials: (S) Pattern block shapes, 2 right triangles (Template 2), 3-piece square puzzle (Template 3, cut into 3 pieces), puzzle template (Template 4)

1. (Give the student two right triangles.) Use these triangles to make a rectangle.
2. (Give the student the 3-piece paper square puzzle disassembled.) This was a square. Then, I cut it into three pieces. Can you put it together so it makes a square again?
3. (Place the pattern blocks and puzzle template in front of the student.) Use your pattern blocks to complete the puzzle.

What did the student do?	What did the student say?
1.	
2.	
3.	

Module 6: Analyzing, Comparing, and Composing Shapes

End-of-Module Assessment Task

A STORY OF UNITS   K•6

	Class Record Sheet of Rubric Scores: Module 6		
Student Names:	**Topic A:** Building and Drawing Flat and Solid Shapes	**Topic B:** Composing and Decomposing Shapes	**Next Steps:**

Module 6:   Analyzing, Comparing, and Composing Shapes